Roundy & Friends
Book Five

Andres Varela

Illustrations and Graphic Design by Carlos F. González
Co-Producer Germán Hernández
Fourth Edition

Previously...

   <u>Houston</u>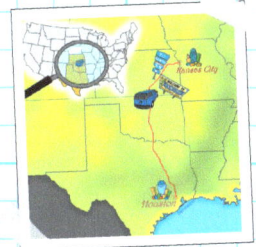

 <u>Kansas City</u>

 <u>Chicago</u>

 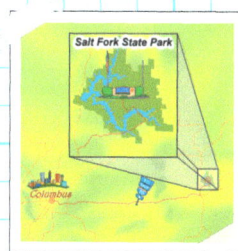  <u>Columbus</u>

The team is heading from Columbus to Washington D.C., the capital of the United States. They take highway I-70 East then I-270 South.

They arrive at an extended stay hotel. Extended stay hotels are places where people typically stay for long periods of time, therefore each room has a full kitchen, small dining room and a living room to make people feel more comfortable while sleeping and living there.

They enter the room and get comfortable, Ben and Gabe check out the bedrooms and Jersey checks out the kitchen, while the others get comfortable in the living room area.

They decide to cook spaghetti for dinner. They all go to the nearest supermarket to get all the ingredients.

They have a list of ingredients they look for including, spaghetti pasta, onions, green and red peppers, mushrooms, vegetable oil and parmesan cheese. The extended stay hotel provides the salt.

Can you help the team count how many of each ingredient they have?

The spaghetti is ready! Jersey and Shorty help serve the plates of spaghetti to the rest of the group. It has been a long day and they need to get ready to go to bed, so they can wake up early in the morning to head over to the airport to pick up Danielle and Marie who are flying from Canada to join this adventure.

Early in the morning they go to the Washington D.C. Dulles Airport, which is located about 26 miles west of downtown Washington D.C.

They park the car and head over to the baggage claim area, which is where people go when they need to claim their checked luggage after their flights. They find Danielle and Marie!

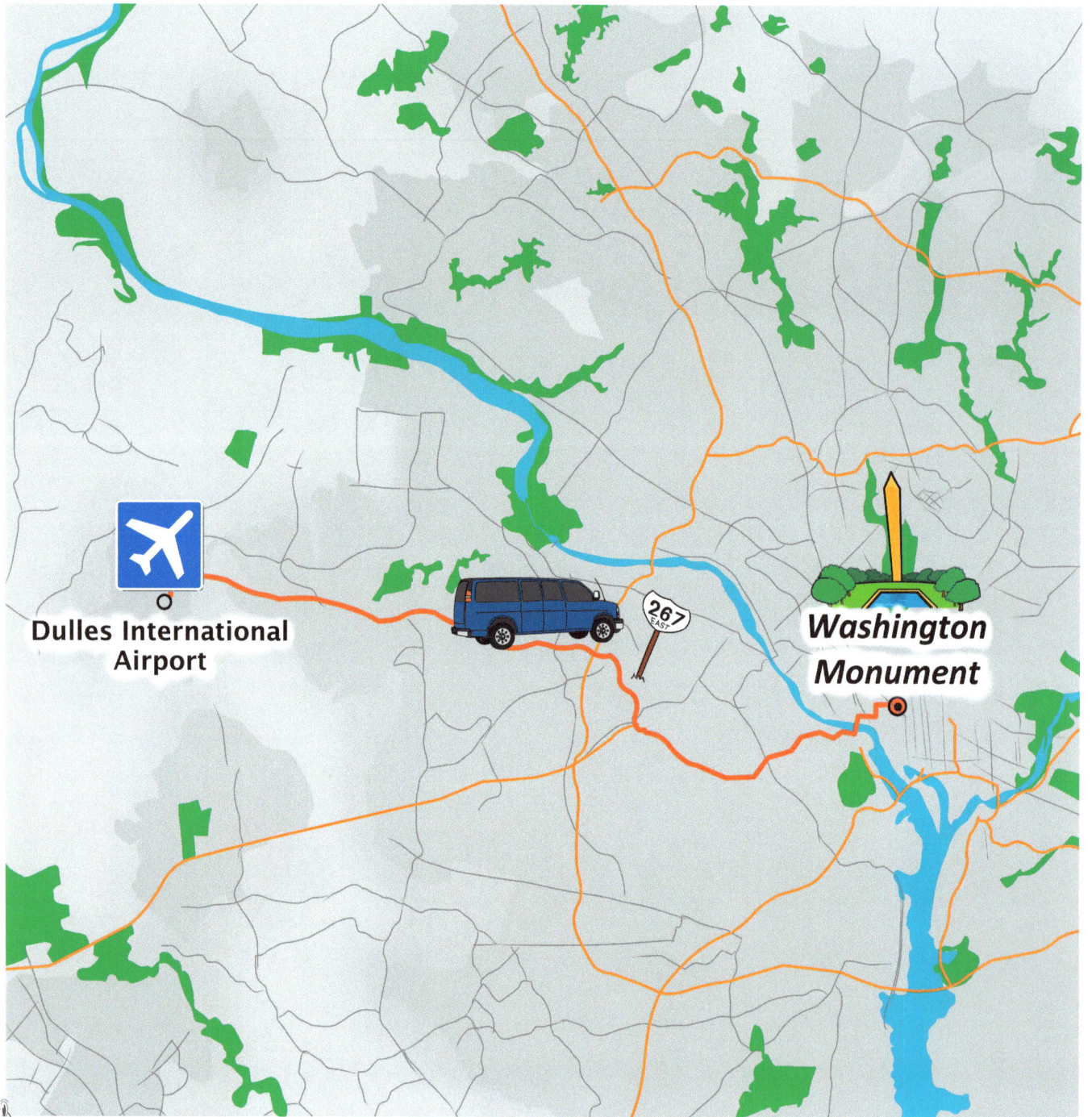

Dulles International
Airport

267
EAST

*Washington
Monument*

The ride back to the city takes about 40 minutes with traffic. They have a busy day of sightseeing and visiting many very famous places in the country's capital city.

The first stop is the Washington Monument, built in 1885, and at the time it was the tallest structure in the world.

The monument is a stone pillar, also called an obelisk. The obelisk is over 555 feet or 169 meters tall. The monument is dedicated to George Washington, the first American President.

The water in front of the obelisk is called the Lincoln Memorial Reflective Pool.

The Lincoln Memorial is a monument in honor of the 16th president of the United States, Abraham Lincoln. It's located across from the obelisk and the Lincoln Memorial Reflective Pool. Construction was completed in 1922.

Above the columns or pillars are the names of the 36 States that were part of the Union at the time of Lincoln's death and the dates on which they entered the Union.

In the top portion of the structure are the names of the 48 States part of the Union when this monument was dedicated in 1922.

After visiting Lincoln's Memorial the team heads over to the White House. They see a helicopter taking off. Teo tells the team "That is Marine One." which is the call sign designated to the helicopter carrying the president of the United States. Marine two is the call sign for the helicopter carrying the Vice-President.

The White House is the residence and official work place of the President of the United States. It is the same residence used for all presidents since 1800, the first president to live there was John Adams. The ZIP code where the house is located is 20500.

The White House

The Pentagon

"What is a ZIP code?" asks Roundy.

Teo responds "ZIP codes are a system to designate postal codes in the United States. It was created in 1963. ZIP means Zone Improvement Plan and was chosen to suggest that the mail travels more efficiently and therefore more quickly, or zipping along, when the senders use the code in the postal address".

They leave the White House and start walking over to The Pentagon.

The walk over the Potomac River is very beautiful. It is a bright day for a walk.

The Pentagon is the headquarters of the United States Department of Defense. The building was finished and dedicated in 1943. The structure has five sides and on any given work day it employs over 26,000 people, between military personnel and civilians.

Smithsonian American Art Museum

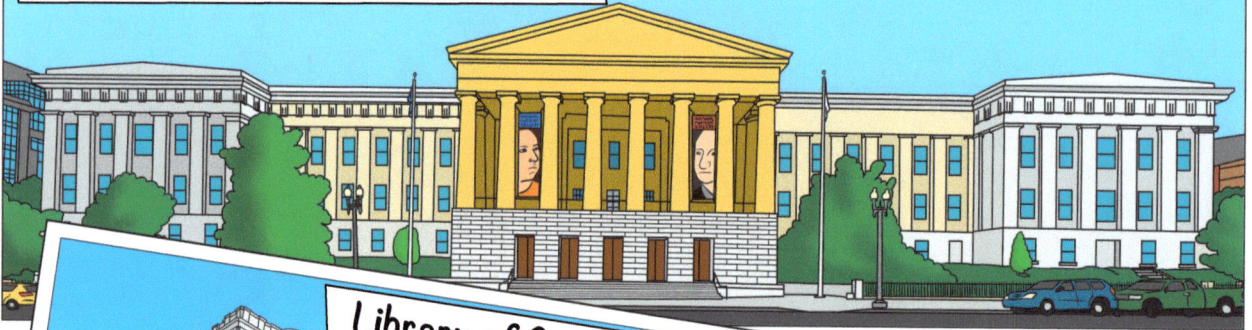

Library of Congress

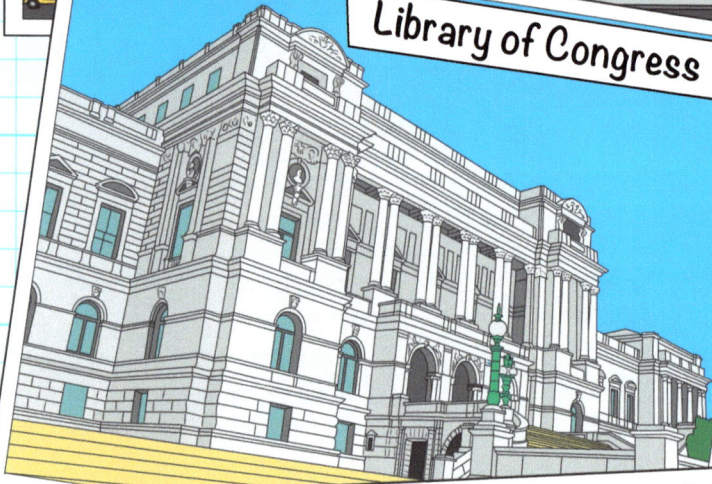

The Luce Foundation Center for American Art

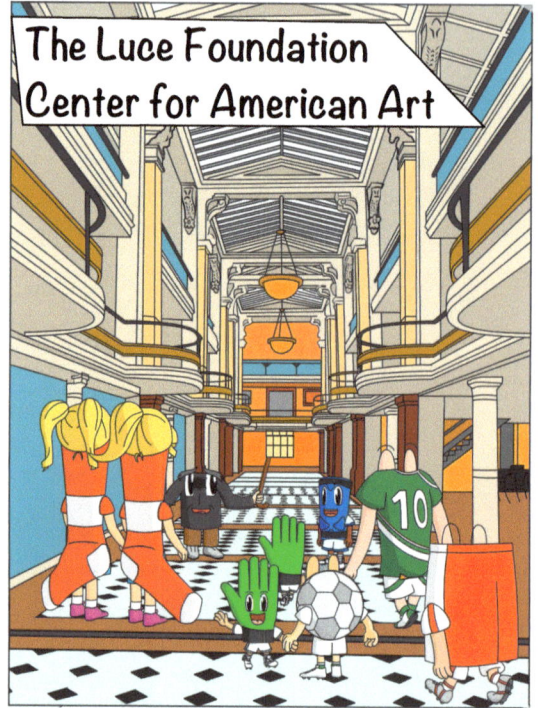

Thomas Jefferson Library

The rest of the day is very busy for our friends. They go to the Library of Congress which is the national library of the United States, as well as the research library that officially serves the United States Congress. They have several exhibitions, like the recreation of the Thomas Jefferson Library. They also visit the Smithsonian American Art Museum, which contains one of the largest collections of art in the world, from the colonial period to the present. On the third floor of the Museum is the Luce Foundation Center of American Art which was opened in the year 2000 and contains over 3,300 objects on display.

The group takes time to reflect on this trip to Washington DC. They go to a park where they relax and talk about everything they learned!

They visited Dulles International Airport, Lincoln Memorial, the Washington Monument, the White House, the Pentagon, the Library of Congress and the Smithsonian American Art Museum. They also visited the stadium where the soccer team from Washington plays is called RFK stadium, there are also baseball games played there.

A lot of the Museums and exhibitions in Washington D.C. are free to the public, you should come and visit them!

Come back and read the next story, we're heading to Philadelphia.....